A BETTER MODEL OF THE SOLAR SYSTEM

AND

WHY THE UNIVERSE IS NOT EXPANDING

Albert W. McKinney III
2019 April 22

DEDICATION

This book is dedicated to Phyllis McKinney, my wife of 68 years.

Copyright © 2019 by Albert W. McKinney III

Library of Congress Cataloging-in-Publication Data
McKinney, Albert William III (1929-)
    A Better Model of the Solar System and Why the Universe is Not Expanding.
    CreateSpace Independent Publishing Platform,
        North Charleston, South Carolina
    ISBN: 9781095580165

TABLE OF CONTENTS

PAGE	ITEM	CONTENT
v		PREFACE
1		INTRODUCTION
3	PART 1	THE ORIGIN OF GRAVITY
3		What is Light? (preliminary)
4		Electrons and Protons Vibrate
5		What Happens in a Single Vibration?
6		What Generates the Flood?
6		What is Light? (final)
6		Summary of Part 1
7		APPENDIX TO PART 1
7		Justification for Case (2)
7		Justification for Case (3)
8		The Strong Nuclear Force
9	PART 2	HOW GRAVITY WORKS
11	PART 3	HOW GRAVITY LEADS TO PERIHELION PRECESSION
11		Preliminary Discussion
15		APPENDIX TO PART 3
15		Proofs of equations (3-1) and (3-2)
15		Differential Equations of Motion
16		*Vis Viva* Integral
17		The Main Differential Equation
19		Approximation 1
19		Approximation 2
23	PART 4	CONSEQUENCES OF A CENTURIES-OLD ERROR
23		The Error
24		The Consequences
25	PART 5	CONJECTURE ON THE RATE AT WHICH A PROBE LOSES MASS
27		Working Backwards

29	PART 6	EPILOGUE
29		General Relativity
31	ADDENDUM	BLACK HOLES?

PREFACE

In 1687, Newton published his law of gravity, and showed how gravity explained the movements of the planets.

In 1857, Le Verrier discovered that the movement of the perihelion of Mercury did not quite follow Newton's law. The resulting difference is called the precession of the perihelion. Le Verrier concluded that Newton's law was not quite exact.

In 1916, Einstein developed general relativity, and showed how it allowed for the measurement of the precession of the perihelion of each planet, apparently solving the problem discovered by Le Verrier. He did so by making several assumptions, such as "space-time is curved."

The present book approaches the situation in a completely different way. It makes certain assumptions about the interactions between particles. These assumptions are mostly based on observed properties of matter.

One interesting conclusion is that gravity can be blocked by the matter in a star.

The result is a model of how gravity works. This then yields a model of the solar system in which the precession of the perihelion of each planet is almost exactly explained. The model includes equations of the orbits of each planet. These equations describe the main causes of the planets' orbits, but lack the effects of other planets on a given planet's orbit (hence the use of the word *almost* above).

Once these matters are dealt with, attention is given to the way in which light is transferred through space. Fritz Zwicky discovered that light loses mass and energy as it travels through space. This implies that the measurements of Edwin Hubble are biased, and that his results are always larger than the correct values. This invalidates the idea that the universe is expanding. Consequently, there is no longer any reason to believe that there was ever a Big Bang, nor that there is any need to believe in the existence of dark matter or dark energy.

INTRODUCTION

This introduction contains a summary of the material found in the rest of the book.

A key to understanding gravity is to understand the transmission of light. For both of these phenomena start from the same physical event: the vibration of an electron or a proton.

Vibration? Well, it is known that each of these particles has a frequency: The frequency of an electron at rest is 1.2×10^{20} cycles per second, and that of a proton at rest is 2.3×10^{23} cycles per second. By special relativity, these frequencies increase if the particles are moving.

A frequency implies some kind of repeated motion. What is this motion?

ASSERTION:

At the beginning of a cycle, a particle (electron or proton) emits a *probe*. The probe flies out in some direction and then returns to its particle. On its outward journey, the probe may

(1) hit nothing, run out of time, and return to its particle at the end of the cycle; or

(2) hit one or more other probes, then run out of time, and return to its particle at the end of the cycle; or

(3) hit zero or more other probes, then hit another particle, then return to its own particle (thus ending the cycle).

In case (3), a study of the deuteron shows that after hitting another particle, the probe then returns to its own particle, thus ending that cycle. The result is to transfer an electromagnetic impulse to the target particle, (for example, an impulse which causes the emission of light). After this transfer is complete, the current cycle is cut short, and the probe returns to its particle to end the cycle. This case is the sole cause of electromagnetism.

Case (2) is implied by the observed phenomenon of the path of light from a star being bent as the probe carrying the light passes by the Sun. That is, a probe from a star hits a probe from the Sun. The interaction bends the path of the star probe toward the Sun, but that probe continues on the new path (else we would not see this deviation at all!). The interaction also bends the path of the probe from the Sun, but this is not observable.

In case (2), one effect of the hit is to yield a pulse of gravity which attracts both particles to the point of the hit. A second effect is to modify the paths of the two probes and send them on in slightly new directions. This case is the sole cause of gravity.

Case (1) is the null case; nothing happens.

Why is it reasonable to state that case (3) is the sole cause of electromagnetism and case (2) is the sole cause of gravity? It is because in this theory, those are the only ways in which something can happen, and there are only two things that can happen: either an electromagnetic connection or a gravitational connection. The only other case is case (1), which does not yield any force at all.

For an object such as a planet or an asteroid, the probes emitted by particles within the object fly so fast that the chance of case (2) or (3) happening within the object is negligible. This is not so for a star. For example, a probe emitted by an electron on one side of the Sun in the direction of the opposite side of the Sun stands a large chance of hitting a particle within the Sun (a case (3) situation). This prevents the probe from leaving the Sun on that cycle. Consequently, that probe cannot encounter a case (2) or a case (3) situation on any planet or asteroid outside the Sun.

It turns out that there are a huge number of these situations in the Sun. As a result, to a planet or asteroid, the apparent center of mass of the Sun is offset from its true center of mass by about 4,375 meters toward that planet or asteroid.

Thus the orbit of a planet or asteroid can be found using Newton's law of gravity based on a separation r equal to the distance between that object and the true center of mass of the Sun **minus** 4,375 meters.

Of course, there is a corresponding need to use a reduced mass for the Sun, but that reduced mass is so close to the measured solar mass that the difference is less than the least significant digit in the measured solar mass, so that difference is negligible.

The orbits resulting from this procedure have the property that they yield the observed value of the precession of the perihelion of each object.

A further consequence of Zwicky's discovery is this: It has become apparent that Edwin Hubble used an incorrect formula to evaluate the velocities of stars. The formula he used is

$$z = \frac{\lambda_{rec} - \lambda_{em}}{\lambda_{em}},$$

where λ_{rec} is the wavelength of the light received on Earth, and λ_{em} is the presumed wavelength of the light emitted at the star. Hubble assumed (as had Einstein and Newton) that the wavelength of the light received on Earth, λ_{rec}, is the same as the wavelength emitted by the star.

But Zwicky's observation implies that that light loses mass and energy during the passage from star to Earth, and consequently gains in wavelength. Thus λ_{rec} needs to be corrected for that gain. Unfortunately, such a correction requires knowledge of λ_{rec}, and such values have not (to my knowledge) been published. Also, a formula for such a correction has not been published. A conjecture in Part 5 below may serve the purpose.

PART 1: THE ORIGIN OF GRAVITY

ASSERTION:

On Earth, we live in a flood of light.

How so?

Consider anyone with reasonable eyesight; for example, you who reads this. Look around you. You see whatever exists surrounding you.

How is it that you see those things?

It is because almost every particle on the surface of those things is constantly sending you a stream of information about itself.

Constantly? Well, if it were not a constant stream, that is, if it were intermittent, you would expect it to have gaps in it. It would flicker.

But it does not flicker. Any object in view will provide a steady image of itself for your eyes.

We live with this all our lives, so it does not seem unusual. But consider that there are perhaps billions or trillions (or more) particles on the surfaces of the objects surrounding you. And for each of them to provide steady views to you, they must each be sending you thousands or more bits of information every second. Thus in total, your eyes are receiving perhaps quadrillions of bits of information every second.

That is why I say that on Earth, we live in a flood of light.

This assertion leads to questions: What is light? What causes all those particles to produce the flood of light?

What is Light? (preliminary)

The currently accepted concept of light is that a unit of light is a photon, with a tiny bit of energy but no mass, which flies through space at a speed of 299,792,458 meters per second.

Unfortunately, this concept is wrong. Here is one reason:

Early in the twentieth century. Einstein and others showed that mass and energy are related by the formula $E = mc^2$. That formula has been verified many times. The implication is that mass and energy are merely different aspects of a physical object. Any object which has mass also has energy, and any object which has energy also has mass, and the relation between the two is given by the above formula.

This means that no physical object can consist only of energy without any mass. Hence *photon* is an impossible concept.

To provide an adequate answer about what light is, some further ideas must be explored.

Electrons and Protons Vibrate

Every object that we see is composed of atoms. Atoms are composed of electrons and protons. And electrons and protons have frequencies; they vibrate. How so?

An electron has a rest mass $m_e = 9.109\,382\,91 \times 10^{-31}$ kg, and a proton has a rest mass $1.672\,621\,777 \times 10^{-27}$ kg. By the rule $E = mc^2$, it follows that these particles have rest energies

$$\begin{aligned} E_e &= m_e c^2 \\ &= 9.109\,382\,91 \times 10^{-31} \times (299,792,458)^2 \text{ joules} \\ &= 8.187\,105\,07 \times 10^{-14} \text{ joules} \end{aligned}$$

$$\begin{aligned} E_p &= m_p c^2 \\ &= 1.672\,621\,777 \times 10^{-27} \times (299,792,458)^2 \text{ joules} \\ &= 1.503\,277\,484 \times 10^{-10} \text{ joules} \end{aligned}$$

By the rule $E = h\nu$, where h is Planck's constant and ν is the frequency of the particle, these particles have rest frequencies

$$\begin{aligned} \nu_e &= E_e/h \\ &= 8.18710507 \times 10^{-14} / 6.62606957 \times 10^{-34} \text{ cycles per second} \\ &= 1.235\,589\,965 \times 10^{20} \text{ cycles per second} \end{aligned}$$

$$\begin{aligned} \nu_p &= E_p/h \\ &= 1.503\,277\,484 \times 10^{-10} / 6.62606957 \times 10^{-34} \text{ cycles per second} \\ &= 2.268\,731\,815 \times 10^{23} \text{ cycles per second} \end{aligned}$$

The length of a cycle is of interest. From the above, it follows that an electron cycle c_e is

$$\begin{aligned} c_e &= 1/(1.235\,589\,965 \times 10^{20}) \text{ seconds per cycle} \\ &= 8.093\,299\,787 \times 10^{-21} \text{ seconds per cycle,} \end{aligned}$$

and a proton cycle c_p is

$$c_p = 1/\left(2.268\,731\,815 \times 10^{23}\right) \text{ seconds per cycle}$$
$$= 4.407\,748\,829 \times 10^{-24} \text{ seconds per cycle}$$

As much as is known about electrons and protons, it is curious that we do not know what either particle consists of. For the present purpose, I assume that each particle consists of a substance which I will call *matter*. We know that this matter can be measured in terms of how much mass it has, or how much energy it has.

What Happens in a Single Vibration?

In general, vibration involves movement. In order to reflect observations, the best description of the movement involved in one vibration of an electron (one electron cycle) is the following:

At the beginning of a cycle, the matter in an electron is located at the *core* of the electron (the *position* of the electron). But that matter is moving at a very high rate of speed. As the cycle continues, the matter *unrolls*, acting as a *probe* in some direction. That is, it leaves a fine trace of matter connecting it to its core, and the remainder of the matter rolls on until either (Case (3)) it hits another particle, or (Case (2)) it hits another probe, or (Case (1)) it runs out of matter. In any case, it then rolls back up until it reaches its core, which ends the cycle. A new cycle begins immediately, and the matter unrolls as a probe in a nearly opposite direction. These nearly opposite directions cause the probes from a particle to take virtually all directions from the core over a period of a second.

A similar description applies to a proton.

Case (1) is simply the null rule. The reasoning for choosing Cases (2) and (3) is given below, in the Appendix to Part 1.

When a probe hits another particle, the result is an electromagnetic interaction. This generally involves the transfer of a tiny bit of matter between the probe and the target particle.

When a probe hits another probe, the result is a gravitational interaction. This has two results: The paths of the two probes are slightly changed, and the cores of the two probes are affected by either a small attraction or a small repulsion. **This is where gravity originates.** The cumulative effect of these small attractions and repulsions ends up as a gravitational attraction. For details, see Part 2.

Of course, the above descriptions stem from my own imagination, and have no basis in physical observations. However, I assert that they lead to so many reasonable results that they have a high chance of being accurate.

That a probe loses matter as it unrolls reflects physical observations by Fritz Zwicky, who analyzed data obtained by Edwin Hubble, and noticed that light from a star seemed to lose energy in proportion to the distance it travels from its star ("tired light").

What Generates the Flood?

The flood of light is but a small subset of the gigantic number of probe-probe and probe-particle hits that occur every second. The vast majority of these hits result in no light occurring. Instead, many of them result in the matter being immediately reflected to some other particle. Some of them are merely absorbed by the target particle, thus raising its temperature.

What is Light? (final)

Light is simply the effect which results when a probe hits a particle in an eye (human or animal). Eyes have evolved so they can analyze many things about the hit, such as how intense it is, what frequency it is, and what direction it comes from.

So light is not a substance, not a particle, not a wave, but simply an ephemeral event in the eye. Light does not travel anywhere, hence the speed of light is zero! Of course, the occurrence of light in the eye triggers physical reactions in the body, but light itself is only the trigger.

Summary of Part 1

Electrons and protons vibrate. In one vibration cycle, either type of particle emits a probe. The probe consists of the matter of the particle. This probe unrolls its matter in some direction until either (3) it hits the core of another particle, interacts with it, and then returns to its core, or (2) it hits another probe, interacts with it, thus changing its direction slightly, and continues on, or (1) it runs out of matter and returns to its core.

In case (3), the interaction yields an electromagnetic interaction between the corresponding cores. The interaction can lead to a spark of light in the target core.

In case (2), the interaction yields a gravitational interaction between the corresponding cores.

APPENDIX TO PART 1

Justification for Case (2)

The observed bending of light rays by the Sun, in the terminology of this paper, is as follows. A probe from another star passes near the Sun. As it passes, a probe from a particle within the Sun hits the probe from the star. As a result, the path of the probe from the star is changed to pass closer the Sun.

The implication here is that when two probes meet, the collision results in two tiny attractive forces, one between one particle and the point of the collision, the other between the other particle and the point of the collision. The accumulation of such tiny forces results in gravity. This is the justification for Case (2).

Justification for Case (3)

This justification is based on a property of the deuteron (the nucleus of the deuterium atom). A deuteron consists of a proton and a neutron. A neutron decays into a proton and an electron. Hence for this analysis, we will consider that a deuteron consists of two protons and an electron.

A fundamental property of electrons and protons is that they vibrate, that is, they emit probes. Each such particle emits probes at a frequency which is at a minimum when the particle is at rest, but is greater when the particle is moving. Therefore, it follows that the three particles of the deuteron must be emitting at least as many probes as they would if all three were at rest. Corresponding values are shown in Table 1-A1.

TABLE 1-A1 Number of Probes Emitted per Second by a Particle at Rest (its Frequency)

Particle	Frequency (s^{-1})
Electron	$1.235\,559 \times 10^{20}$
Proton	$2.268\,732 \times 10^{23}$

This table shows that the three particles of the deuteron must emit at least

$$1.235\,559 \times 10^{20} + 2 \times 2.268\,732 \times 10^{23} = 4.538\,700 \times 10^{23} \equiv P_1$$

probes per second, which corresponds to a mass of $3.346\,155 \times 10^{-27}$ kg. (This follows from the special relativity formula $mc^2 = h\nu$, where m is the particle mass, c is the velocity of light, h is Planck's constant, and ν is the particle frequency.) And it is quite likely that these nuclear constituents are moving, so that their frequencies are greater than the rest frequencies.

However, the measured mass of the deuteron is only $3.343\,583 \times 10^{-27}$ kg., which corresponds to $4.535\,211 \times 10^{23} \equiv P_2$ probes per second. Thus the deuteron would seem to have a deficiency of at least $P_1 - P_2 = 3.489 \times 10^{20}$ probes per second. How can this deficiency be explained?

To set the stage for an explanation, assume that the particles in the deuteron emit a total of N_e probes per second, but that N_a of them are absorbed by one of the other nuclear particles, and the remainder, $N_m = N_e - N_a$, miss the other nuclear particles. Assume further that the ones that miss are the only ones that can be observed, and that they account for the measured mass of the deuteron. In this case,

$$N_m = P_2 = 4.535\,211 \times 10^{23}$$
$$N_e \geq P_1 = 4.538\,700 \times 10^{23},$$

hence

$$N_a = N_e - 4.535\,211 \times 10^{23} \geq 3.489 \times 10^{20}.$$

It seems reasonable to assume that the N_a absorbed probes account for the binding energy of the deuteron, which is about 2.224 52 MeV. This is equivalent to $3.965\,566\,734 \times 10^{-30}$ kilograms, and to a frequency of $5.378\,865\,406 \times 10^{20}$ probes per second. That is, $N_a = 5.378\,865\,406 \times 10^{20}$.

This implies that sometimes a probe emitted by a nuclear particle will hit another particle in the same nucleus and will not proceed outside of the nucleus (a Case (3) situation). It was for this reason that it was stated earlier that when a probe hits another particle, it stops there and does not continue on its outward journey.

A further implication is that the measured mass of an atom is based on the probes which emerge from the atom, but does not count the probes which connect with other particles in the atom. Therefore, for any atom with more than a single particle in its nucleus, its actual mass is greater than its measured mass, and the velocities of the particles in the nucleus are greater than would be suggested by the measured mass of the atom.

The Strong Nuclear Force

Above it was shown that electromagnetic force and gravitation are explainable in terms of probes. The strong nuclear force can also be thus explained. The key lies in atoms with nuclei having more than 1 component. In such atoms, the probe of a nuclear constituent sometimes hits another nuclear constituent as it begins its cycle, and thus never leaves the nucleus on that cycle. Such hits account for the strong nuclear force.

PART 2: HOW GRAVITY WORKS

In Part 1 it was mentioned that gravity is a consequence of two probes hitting. There are a couple of physical results that lead to this conclusion.

First, it is clear that a light ray from a star is bent towards the Sun when it passes close to the Sun. Translating this to the terminology of this paper, the light ray passing the Sun is merely a probe from a particle in a distant star. The cause of its path being deflected toward the Sun is that it is hit by a probe from a particle in the Sun. This causes the first probe to change its direction slightly toward the Sun.

Since this reaction always takes place, that is, the path of a light ray passing near the Sun always changes direction toward the Sun, it follows that when two probes hit, they always interact attractively.

But this presents a small problem: If the hit occurs at a point in space which is not between the two particles, this would lead to a negative force between the two particles. This is resolved as follows.

Consider two cores that are emitting probes. Some of these probes will connect with the other core. Others will not connect with either the other core or its probe. But some of them will connect with the probe of the other core. These probe-probe connections lead to gravity in the following way.

While such connections can occur almost anywhere, the vast majority of them occur in two clusters, one centered around each of the cores from which the two probes are emitted. When such a connection occurs, it causes a force to be exerted on each core, pulling it toward the location of the *connection*.

The connections in one cluster can be almost anywhere around the core, except for a tiny cone extending from the core out in a direction opposite to the other core. The reason no connection can occur in that cone is quite simple: Take core 2 for example, as shown in Figure 2-1. To get to a connection within the cone on the right side of particle 2, the probe from core 1 would have to pass close enough to core 2 to form a probe-core connection. This would prevent core 1's probe from forming any other connections on that cycle.

Figure 2-1 Connection-free Zone of a Particle

The zone to the right of core 2 will be called the no connection zone (or *nocon* zone) of particle 2 with respect to particle 1.

Thus, for a pair of particles, each particle has a nocon zone with respect to the other particle. Probe-probe connections can occur anywhere around a particle except within its nocon zone.

Using the above figure as an example, consider a vertical plane through particle 2. Some probe-probe connections in the cluster about particle 2 will lie on the same side of that plane as particle 1. Each such connection will cause particle 2 to experience a slight attraction to particle 1, as indicated in Figure 2-2.

Figure 2-2 When a Connection Occurs to the Left of Particle 2

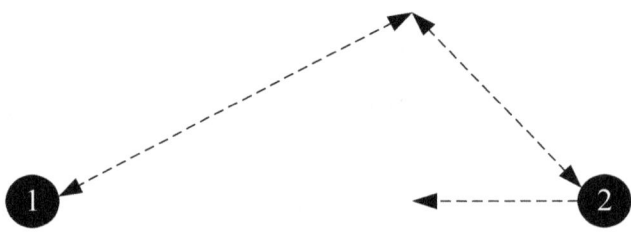

However, other probe-probe connections in that cluster will lie on the other side of that plane, and will cause particle 2 to experience a slight repulsion from particle 1, as indicated in Figure 2-3.

Figure 2-3 When a Connection Occurs to the Right of Particle 2

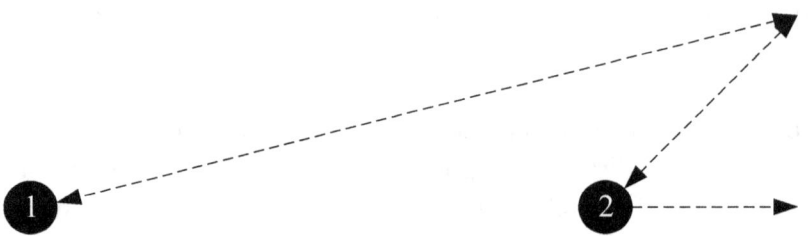

If nocon zones did not exist, then the number of probe-probe connections leading to attractive force on particle 2 would be the same as those leading to repulsive force on particle 2, and the net result would be zero force on particle 2. Thus gravity would not exist! However, the existence of the nocon zone of particle 2 means that there will be slightly more connections leading to attractive force on particle 2 than those leading to repulsive force.

And this slight discrepancy results in gravity!

Note: The forces involved in probe-probe connections are far larger than gravity. Gravity is an average of gazillions of probe-probe connections, which do not quite average out to zero. There is generally no single interaction between particles which results in a force equal to gravity.

PART 3: HOW GRAVITY LEADS TO PERIHELION PRECESSION

Consider a particle within a star. At the beginning of a cycle, the particle emits a probe. As mentioned above, the probe goes out from the core of the particle until either (3) it hits another particle, or (2) it hits another probe, or (1) it runs out of matter. For such a probe, there are a vast number of other particles that it must pass on its way toward the surface of the star. Small wonder that the probe will sometimes hit another particle within the star on its way out. This will cause the probe to stop and return to its core. In other words, there is a chance that a probe from a particle within a star will not be able to get outside the star on some cycles.

If so, then that probe cannot make gravitational contact with any object outside the star on such cycles. To such an object, the result is that it does not experience the full mass of the star. It is as if a portion of the star (typically on the side opposite to that object) is invisible to that object. Consequently, that object perceives the center of mass of the star as being offset toward that object from its true center of mass (a *gravitational offset*).

This may seem as a trivial amount. It is not. Take the Sun for example. Experimental evidence shows that the Sun's gravitational offset is about 4,375 meters.

What is this evidence? It is the precession of the perihelia of the planets, asteroids and comets orbiting the Sun. These objects actually orbit the *apparent* center of mass of the Sun, not its *true* center of mass. A bit of celestial mechanics shows that the amount of the precession of each perihelion is directly related to the Sun's gravitational offset. And if the Sun's gravitational offset is 4,375 meters, then the precessions of the various planets are just what has been measured by observations.

ASSERTION:

The cause of the precessions of the perihelia of the planets, asteroids and comets that orbit the Sun is a very simple result of the way that gravity works: *Gravity is blocked by a mass nearly the mass of the Sun. Consequently, the apparent center of mass of the Sun is offset from its true center of mass by about 4,375 meters.*

The proof of this assertion will be given later in this part. Before doing that, it is useful to discuss the situation.

Preliminary Discussion

Consider some of the consequences which follow if the assertion is true.

An observer on Earth cannot then measure the entire mass of the Sun. The reason for this is that there is a part of the Sun, say P, the gravitational force of which is blocked by the remaining mass of the Sun. (Obviously P is on the opposite side of the Sun from the observer.) Thus to the observer, it appears that the center of mass of the Sun is offset toward the observer from the true center of mass of the Sun.

The point is that if the assertion is true, then every planet and asteroid which orbits the Sun is subject to the gravitational force of only a (very large) part of the Sun, and that force is centered on an apparent center of mass of the Sun which is offset from the Sun's true center of mass.

Using standard celestial mechanics in a Euclidean universe, one can derive two useful equations. .The first is the equation of motion of such a planet or asteroid (the object):

$$(3\text{-}1) \quad p = \frac{1}{A\cos\left(\sqrt{1-2Lq}\,\theta - \omega\right) + \frac{L}{1-2Lq}}$$

Note that this equation does not take into account the influence of other planets or asteroids, though these are always very small effects. The second equation is the value of the precession of the object, in arc seconds per orbit.

$$(3\text{-}2) \quad 360 \times 3600 \left(\frac{1}{\sqrt{1-2Lq}} - 1\right) \quad \text{in arc seconds per orbit.}$$

In these equations,

> A and ω are constants specific to the object;
> p is the distance of the object from the apparent center of mass of the Sun when the object is at a point θ in its orbit;
> q is the value of the gravitational offset of the Sun;

and

$$L = \frac{G(M_1 + M_2)T^2}{4\pi^2 a^4 (1-e^2)},$$

where

> G is the gravitational constant;
> M_1 is the measured mass of the Sun;
> M_2 is the mass of the object;
> T is the time of one orbit;
> a is the semimajor axis of the ellipse;
> e is the eccentricity of the ellipse.

The last three values make use of the fact that the curve given by equation (3-1) is always very close to an ellipse.

For the calculations below, three constants will be needed:

gravitational constant	$6.674\,08 \times 10^{-11}$ m^3 kg^{-1} s^{-2}
measured mass of the Sun	$1.998\,4 \times 10^{30}$ kg
seconds per Julian century	3,155,814,977 s / Jc

The next step is to calculate the value of L for each planet. (Hereafter, the asteroid 1566 Icarus will be included with the planets.)

Planet	Mass of the planet (kg)	Time of one orbit (sec)	Semimajor axis (m)	Eccentricity	L
Mercury	$3.302\,200 \times 10^{23}$	7,600,530	57,909,100,000	0.205 630	$1.803\,01 \times 10^{-11}$
Venus	$4.867\,600 \times 10^{24}$	19,414,166	108,208,000,000	0.006 700	$9.241\,79 \times 10^{-12}$
Earth	$5.972\,190 \times 10^{24}$	31,558,150	149,598,261,000	0.016 711	$6.686\,13 \times 10^{-12}$
Mars	$6.418\,500 \times 10^{23}$	59,354,294	227,939,100,000	0.093 320	$4.425\,50 \times 10^{-12}$
1566 Icarus	$2.900\,000 \times 10^{12}$	35,316,850	161,252,061,359	0.826 838	$1.960\,31 \times 10^{-11}$
Jupiter	$1.898\,600 \times 10^{27}$	374,335,776	778,547,200,000	0.048 775	$1.286\,37 \times 10^{-12}$
Saturn	$5.684\,600 \times 10^{26}$	929,596,608	1,433,149,370,000	0.055 723	$6.909\,29 \times 10^{-13}$
Uranus	$8.681\,000 \times 10^{25}$	2,651,486,400	2,875,031,718,261	0.046381	$3.483\,99 \times 10^{-13}$
Neptune	$1.024\,300 \times 10^{26}$	5,200,418,592	4,503,443,661,000	0.011 214	$2.210\,60 \times 10^{-13}$

It would seem that the measurement of the precession of Mercury's perihelion is perhaps the most accurate of such measurements; it is 42.4446 arc seconds per Julian century. Hence the next steps involve using the value of L for Mercury with equation (3-2) to see what value of q comes closest to that measured value. That value turns out to be $q = 4,374.724$ meters.

Using that value, the precession of the other planets are as shown in the following table. For historical reasons, the values given by general relativity are shown to the right of the correct values, and the percentage difference for the general relativity values are to the right of those.

Planet	Precession of the Perihelion (arc seconds per Julian century)	Corresponding General Relativity Values (arc seconds per Julian century)	Fractional Error in General Relativity Values
Mercury	42.4446	42.98	1.013
Venus	8.51735	8.619128	1.012
Earth	3.790798	3.836096	1.012
Mars	1.334068	1.350009	1.012
1566 Icarus	9.931404	10.05008	1.012
Jupiter	0.061486	0.06222	1.012
Saturn	0.013299	0.013458	1.012
Uranus	0.001975	0.001999	1.012
Neptune	0.000761	0.00077	1.012

APPENDIX TO PART 3

Proofs of Equations (3-1) and (3-2)

Differential Equations of Motion

The key idea which explains the shift in the perihelia of planets is this: When two bodies are in orbit around each other, two of the things which determine their orbits are not their centers of mass, but instead their *apparent* centers of mass. They sense each other's presence gravitationally, but each senses the center of mass of the other at a point possibly offset from its true center of mass. These offsets determine their mutual orbits.

Suppose two objects of gravitational masses M_1 and M_2 move in orbits around each other. Denote the positions of their centers of mass by the vectors $\mathbf{P}_1(t)$ and $\mathbf{P}_2(t)$. Let $\mathbf{P} = \mathbf{P}_2 - \mathbf{P}_1$, let $p = |\mathbf{P}|$, and let q_1 and q_2 be the sizes of the gravitational offsets of the two objects. Let $\mathbf{S}_1$ be the point between $\mathbf{P}_1(t)$ and $\mathbf{P}_2(t)$ which is a distance of q_1 from $\mathbf{P}_1(t)$, and let $\mathbf{S}_2$ be the point between $\mathbf{P}_1(t)$ and $\mathbf{P}_2(t)$ which is a distance of q_2 from $\mathbf{P}_2(t)$.

These assumptions are made:

A3.1 Each of the two objects acts as a rigid body.
A3.2 $\mathbf{P}_1$, $\mathbf{S}_1$, $\mathbf{S}_2$, and $\mathbf{P}_2$ lie on a straight line in that order.
A3.3 The offset distances q_1 and q_2 do not change with time.

Let $q = q_1 + q_2$.

The gravitational attraction between the two objects is applied at the two points $\mathbf{S}_1$ and $\mathbf{S}_2$. By the above assumptions, the distance between them is equal to $p - q$. Let $\mathbf{F}_i$ denote the gravitational force on object i; then (by analogy with Newton's law)

(3-A1) $$\mathbf{F}_1 = \frac{GM_1M_2}{(p-q)^2} \frac{\mathbf{P}}{p}$$

and $\mathbf{F}_2 = -\mathbf{F}_1$.

By the first and second assumptions, the forces act on the gravitational masses, and the force vectors are applied at the apparent centers of mass. (One result is that the force on the apparent center of mass of the Sun contributes to the rotation of the Sun.) Hence, the differential equations of motion are

(3-A2) $$M_i \mathbf{P}_i'' = \mathbf{F}_i,$$

where primes denote time derivatives. Since $\mathbf{F}_2 = -\mathbf{F}_1$,

$$M_1 \mathbf{P}_1'' + M_2 \mathbf{P}_2'' = 0,$$

and so the center of mass of the entire system:

$$\frac{M_1 \mathbf{P}_1 + M_2 \mathbf{P}_2}{M_1 + M_2}$$

moves in a straight line.

Now consider the differential equation for $\mathbf{P}$. Since $\mathbf{P} = \mathbf{P}_2 - \mathbf{P}_1$, it follows from equation (3-A2) that

(3-A3) $$\mathbf{P}'' = \mathbf{P}_2'' - \mathbf{P}_1'' = \frac{\mathbf{F}_2}{M_2} - \frac{\mathbf{F}_1}{M_1} = \frac{-K\mathbf{P}}{p(p-q)^2},$$

where

$$K = G(M_1 + M_2).$$

Taking the vector product of $\mathbf{P}$ with the first and last sides of equation (3-A3), it is found that $\mathbf{P} \times \mathbf{P}'' = 0$, which can be integrated directly to yield the fact that $\mathbf{P} \times \mathbf{P}'$ equals a constant vector. Thus, the orbits lie in a plane perpendicular to that vector. Take that plane as the x-y plane, and let the origin be at $\mathbf{P}_1$.

Vis Viva Integral

Put equation (3-A3) in rectangular coordinates:

$$x'' = \frac{-Kx}{p(p-q)^2},$$

$$y'' = \frac{-Ky}{p(p-q)^2}.$$

`Multiply the first equation by $2x'$, the second by $2y'$, and add:

(3-A4) $$2x'x'' + 2y'y'' = \frac{-2K(xx' + yy')}{p(p-q)^2}.$$

But by definition, the square of the velocity, v^2, is given by

$$v^2 = x'^2 + y'^2,$$

and so the left side of (3-A4) equals $(v^2)'$. Also, $p = \sqrt{x^2 + y^2}$ and q is a constant. Hence,

$$(p-q)' = p' = \frac{2xx' + 2yy'}{2\sqrt{x^2 + y^2}} = \frac{xx' + yy'}{p},$$

and so the right side of (3-A4) is equal to

$$\frac{-2K(p-q)'}{(p-q)^2}.$$

Thus, equation (3-A3) can be integrated to obtain

$$v^2 = \frac{2K}{p-q} + c_1$$

for some constant c_1.

The Main Differential Equation

Using polar coordinates, set

$$\mathbf{P} = p(\cos\theta, \sin\theta).$$

Then the second derivative is

$$\mathbf{P}'' = (p'' - p\theta'^2)(\cos\theta, \sin\theta) + (2p'\theta' + p\theta'')(-\sin\theta, \cos\theta),$$

and the differential equation (3-A2) becomes

$$\left(p'' - p\theta'^2 + \frac{K}{(p-q)^2}\right)(\cos\theta, \sin\theta) + (2p'\theta' + p\theta'')(-\sin\theta, \cos\theta) = 0.$$

Since $(\cos\theta, \sin\theta)$ and $(-\sin\theta, \cos\theta)$ are mutually perpendicular unit vectors, the above sum can vanish only if the coefficients of those vectors are each zero; thus

(3-A5) $$p'' - p\theta'^2 + \frac{K}{(p-q)^2} = 0$$

and

$$2p'\theta' + p\theta'' = 0.$$

From the latter equation, it follows that

$$(p^2\theta')' = p(2p'\theta' + p\theta'') = 0,$$

and so for some constant h,

(3-A6) $$p^2\theta' = h.$$

Using this fact in equation (3-A5) yields

(3-A7) $$p'' - \frac{h^2}{p^3} + \frac{K}{(p-q)^2} = 0.$$

Next, use the standard transformation $p = 1/u$, and let the angle θ replace time as the independent variable, with dots indicating derivatives with respect to θ. By the definition of u, along with equation (3-A6), it follows that

$$p' = \frac{-u'}{u^2} = \frac{-\dot{u}\theta'}{u^2} = -\dot{u}p^2\theta' = -h\dot{u}.$$

One more use of equation (3-A6) yields

$$p'' = -h\ddot{u}\theta' = -h^2 u^2 \ddot{u}.$$

Hence, rewriting equation (3-A7) in terms of u rather than p and dividing the result by $-h^2 u^2$ yields

(3-A8) $$\ddot{u} + u = \frac{L}{(1-qu)^2},$$

where

$$L = \frac{K}{h^2}.$$

Approximation 1

Equation (3-A6) represents twice the rate of accumulation of area in polar coordinates. The integral of (3-A6) over one orbit (say for $t = 0$ to T, where T is the time required for the orbit) yields twice the area contained within the orbit. Under the (usually very accurate) approximation that the orbit is an ellipse, that area equals πab, where a and b are the lengths of the semimajor and semiminor axes. Hence, the integral results in the equation

$$2\pi ab = Th,$$

so that

$$h = \frac{2\pi ab}{T} = \frac{2\pi a^2 \sqrt{1-e^2}}{T},$$

where e is the eccentricity of the ellipse. Consequently,

$$L = \frac{K}{h^2} = \frac{G(M_1 + M_2)T^2}{4\pi^2 a^4 (1-e^2)}.$$

Approximation 2

Each offset q_i is a fraction of the radius of the corresponding star or planet, and so the sum of the offsets, q, is much smaller than the separation p between the two objects. Thus the quantity $q/p = qu$ is much smaller than 1, and so for all u of interest, the denominator on the right side of equation (3-A8) can be expanded into a power series which converges rapidly. In fact, an excellent approximation is obtained by dropping all terms of order u^2 and higher. The result is the approximate equation

$$\ddot{u} + u \cong L(1 + 2qu),$$

or

$$\ddot{u} + (1 - 2Lq)u \cong L.$$

The nature of the solution to this equation depends on whether $2Lq$ is smaller or larger than 1. It is not correct to say that $2Lq$ is small merely because qu is small. However, for the solution to be at all meaningful, $2Lq$ must be less than 1; otherwise, the solution would not be periodic, but would either expand or contract exponentially, which is not of interest. Hence for the moment, assume that $2Lq$ is less than 1; it is easy to show that it is positive. This leads to the solution

$$u = A\cos\left(\sqrt{1-2Lq}\,\theta - \omega\right) + \frac{L}{1-2Lq}$$

for constants A and ω, as is easily verified by differentiation. From this, it is possible to solve for p:

(3-A9) $$p = \frac{1}{A\cos\left(\sqrt{1-2Lq}\,\theta - \omega\right) + \frac{L}{1-2Lq}}.$$

For small values of $2Lq$, this equation is very nearly that of an ellipse. Thus equation (3-1), which is the same as equation (3-A9) above, is established.

Recall that p represents the distance of the true center mass of the second object from that of the first object in a coordinate system in which the origin is at the true center of mass of the first object. The minimum value of p occurs when the denominator in equation (3-A9) reaches its maximum, and that happens when the cosine equals 1. Similarly, the maximum value of p occurs when the denominator reaches its minimum, and that happens when the cosine equals -1. Of course, this presumes that the denominator never vanishes for any value of θ. It will be seen below that this is the case.

For an ellipse, the separation p at the time of periastron is equal to $a(1-e)$, and at apastron, p equals $a(1+e)$. Using the fact that equation (3-A9) is very nearly that of an ellipse, form the equations

$$A + \frac{L}{1-2Lq} = \frac{1}{a(1-e)},$$

$$-A + \frac{L}{1-2Lq} = \frac{1}{a(1+e)}.$$

These can be combined to yield

$$\frac{L}{1-2Lq} = \frac{1}{a(1-e^2)}$$

and

$$A = \frac{e}{a(1-e^2)}.$$

The interval between two periastra is found by taking two successive values of θ for which the argument of the cosine is equal to a multiple of 2π. Two such values are θ_1 and θ_2, where $\theta_1 < \theta_2$, and where

$$\sqrt{1-2Lq}\,\theta_1 - \omega = 0$$

and

$$\sqrt{1-2Lq}\,\theta_2 - \omega = 2\pi$$

for some value ω.

The change in θ from one periastron to the next is thus

$$\theta_2 - \theta_1 = \frac{2\pi + \omega}{\sqrt{1-2Lq}} - \frac{\omega}{\sqrt{1-2Lq}}$$

$$= \frac{2\pi}{\sqrt{1-2Lq}}.$$

If this change were equal to 2π, there would be no shift in periastron. Otherwise, the shift in periastron is equal to the above amount minus 2π:

$$2\pi\left(\frac{1}{\sqrt{1-2Lq}} - 1\right) \qquad \text{in radians}$$

$$360\left(\frac{1}{\sqrt{1-2Lq}} - 1\right) \qquad \text{in degrees}$$

(3-A10) $\qquad 360 \times 3600\left(\dfrac{1}{\sqrt{1-2Lq}} - 1\right) \qquad$ in arc seconds.

Thus equation (3-2), which is the same as equation (3-A10) above, is established.

PART 4: CONSEQUENCES OF A CENTURIES-OLD ERROR

The Error

In 1929, the Swiss physicist Fritz Zwicky analyzed data collected by Edwin Hubble. He found that light seemed to lose energy as it traveled through space. But he and others were unable to find any theoretical explanation for this phenomenon.

The reason for this inability turns out to be an erroneous belief in how light is propagated. This belief holds that light travels through space as an individual particle (a photon, in twentieth century terminology). This belief has had a firm grip on physicists for more than 300 years (Isaac Newton clearly believed it to be true). But this belief is wrong! As explained in Part 1 above, all communications between particles take place by means of the vibrations of electrons and protons.

Electrons and protons are not solid lumps. Instead, they have frequencies; they vibrate! Since vibrations imply motion, it is convenient to describe these vibrations as follows:

In one vibration, an electron (or a proton) emits a probe. The probe flies out in some direction until either it hits another particle, or it reaches some maximum distance without hitting another particle. [A study of the deuteron (the nucleus of the deuterium atom) shows that the former outcome does sometimes occur.] In either case, the probe then returns to its particle, thus ending the cycle. On the next cycle, which begins immediately, a probe is emitted in almost the opposite direction.

If a probe does hit another particle on its outward journey, there will be an exchange of mass-energy between the probe and the hit particle, which in turn causes an exchange of mass-energy between the two particles. This is the primary way in which particles communicate.

Zwicky's findings imply that light loses mass-energy as it passes through space. To model this, a formula such as the following seems appropriate:

(1) $\qquad (mass, energy) = (original mass, original energy)(1 - \gamma D)$,

where D is the distance traveled by the probe, and γ is a function to be determined. In other words, the mass and energy of a probe diminish smoothly as the probe travels from its core until it runs out of mass-energy. Formula (1) thus models the behavior predicted for Zwicky's "tired light". It seems likely that γ is a slowly diminishing function of D, but this is merely conjectural at this time.

I made use of data in tables published by Allan Sandage, namely Tables 2, 3, and 4 (Astrophysical Journal V178, 1972). One item in these tables is the redshift z. This number is the relative difference between the wavelength of light emitted by a distant star and the wavelength as observed on Earth:

(A) $$z = \frac{\lambda_{rec} - \lambda_{em}}{\lambda_{em}},$$

where λ_{rec} is the wavelength received at Earth, and λ_{em} is the emitted wavelength. This formula works fine for objects on Earth. Unfortunately, the items in formula (A) correspond to two different times. λ_{rec} corresponds to the time on Earth at which the star was observed. λ_{em} is the value presumed for a star not moving with respect to Earth, and corresponds to the time at which the light left the star. For centuries, λ_{rec} has been thought to be the same whether measured at the star or measured on Earth. But Zwicky's work showed that the mass-energy of the emitted light decays as it travels from the star to Earth, or equivalently, the wavelength increases during that time. Hence there are two wavelengths involved: λ_{recS} and λ_{recE}, where λ_{recS} is the wavelength if it had been measured at the star, and λ_{recE} is the wavelength actually measured on Earth. By Zwicky's work, it follows that $\lambda_{recE} > \lambda_{recS}$. Hence Sandage should have used the formula (A*):

(A*) $$z = \frac{\lambda_{recS} - \lambda_{em}}{\lambda_{em}}$$

instead of formula (A), which is why his values of z are always too high.

Unfortunately, values of λ_{emS} are not easily obtained.

The Consequences

The idea of an expanding universe is supported only by Hubble's erroneous velocities. Lacking that support, and in view of the ridiculous consequences of that idea, it seems reasonable to conclude that the universe is not expanding.

The idea of a Big Bang is supported only by the idea of an expanding universe. Lacking that support, it seems reasonable to conclude that there was never a Big Bang.

Again, lacking the support of an expanding universe, there is no reason to believe that ideas such as dark matter, dark energy, or wormholes in space are anything but plot devices for some great science fiction stories!

PART 5: CONJECTURE ON THE RATE AT WHICH A PROBE LOSES MASS

If indeed the mass of a probe diminishes with distance from the core, it would seem reasonable that the rate of shrinkage would be a smooth function such as constant, linear or quadratic.

One reasonable rate would be proportional to the surface area of a sphere. That is, suppose that the probe tip has the shape of a sphere, and it loses a spherical shell of thickness ε every δ of distance from the core. Suppose that the radius of the probe when it is at its core is the classical electron radius $r_e = 2.817\,940\,322\,7 \times 10^{-15}$ meters.

Then the initial mass of the probe tip will be the mass of an electron, $9.109\,382\,91 \times 10^{-31}$ kilograms, which is also equal to $\frac{4}{3}\pi r_e^3 \mu$, where μ is the density of matter. Hence

$$\mu \approx \frac{9.109\,382\,91 \times 10^{-31}}{\frac{4}{3}\pi r_e^3} = 9.718\,628\,335\,856 \times 10^{12} \text{ kilograms per cubic meter.}$$

Assume that the rate of loss of mass is proportional to the surface area:

(1) $$\frac{\mu dV}{dt} = -\alpha S,$$

where α is a constant factor to be determined. Since

$$V = \tfrac{4}{3}\pi r^3$$
$$S = 4\pi r^2,$$

then (1) becomes

$$\frac{\mu d}{dt}\left(\tfrac{4}{3}\pi r^3\right) = -\alpha 4\pi r^2.$$

or

$$\mu 4\pi r^2 \frac{dr}{dt} = -\alpha 4\pi r^2$$

or

$$\frac{dr}{dt} = -\frac{\alpha}{\mu},$$

hence

$$r = -\frac{\alpha t}{\mu} + c_1,$$

where c_1 is a constant to be determined. But when $t = 0$, $r = r_e$; hence $c_1 = r_e$, and so

$$r = r_e - \frac{\alpha t}{\mu}.$$

Then the equation for the mass becomes

$$\mu V = \tfrac{4}{3}\pi\mu\left(r_e - \frac{\alpha t}{\mu}\right)^3.$$

Now the wavelength λ is equal to $\frac{ch}{E}$, so

$$\lambda = \frac{ch}{E} = \frac{ch}{mc^2} = \frac{h}{mc} = \frac{h}{\tfrac{4}{3}\pi\mu c\left(r_e - \frac{\alpha t}{\mu}\right)^3}$$

or

$$\lambda = \frac{3h\mu^2}{4\pi c\left(\mu r_e - \alpha t\right)^3}.$$

On an electron cycle, the elapsed time t is equal to the distance traveled D divided by the speed of light c:

$$t = D/c$$

Thus the factor $\mu r_e - \alpha t$ is equal to $\mu r_e - \frac{\alpha D}{c}$. This factor goes to zero when $D = \frac{c\mu r_e}{\alpha} \equiv D_{max}$. Thus D_{max} is the maximum distance that an electron probe can travel. In a previous book of mine (Physics Unified by Impulse Theory, 2011 February 15), that distance was estimated to be $1.867\,3\times10^{26}$ meters. Hence

$$\alpha \approx \frac{(299\,792\,458)(9.718\,628\,335\,856\times 10^{12})(2.817\,940\,322\,7\times 10^{-15})}{1.867\,3\times 10^{26}}$$

$$\approx 4.396\,868\times 10^{-20},$$

and so

$$\lambda = \frac{3h\mu^2}{4\pi c\left(\mu r_e - \dfrac{\alpha D}{c}\right)^3} = \frac{3h\mu^2 c^2}{4\pi(c\mu r_e - \alpha D)^3},$$

where the only variables are λ and D.

Working Backwards

The aim of this section is to take a value of λ_{rec}, a value measured at Earth, and λ_{em}, an ideal value assumed at the star, and find an estimated value λ_{org} which is the value λ_{rec} would have had when it left the star. This will enable us to form the value

$$z^* = \frac{\lambda_{org} - \lambda_{em}}{\lambda_{em}}.$$

Then z^* is the correct ratio for the relative velocity of the star with respect to Earth.

First calculate

$$\lambda^* = \frac{3h\mu^2 c^2}{4\pi(c\mu r_e - \alpha D)^3},$$

using the value of D for that star. Then subtract λ^* from λ_{rec}:

$$\lambda_{rec} - \lambda^* = \lambda_{rec} - \frac{3h\mu^2 c^2}{4\pi(c\mu r_e - \alpha D)^3}.$$

Finally, calculate

$$z^* = \frac{\lambda_{rec} - \lambda^* - \lambda_{em}}{\lambda_{em}}.$$

PART 6: EPILOGUE

General Relativity

One may wonder how this book relates to general relativity. The answer is that it replaces general relativity. For general relativity is an approximation to reality. It was an amazing feat that Einstein could develop such a good approximation. And for some purposes, general relativity is still a useful tool.

On the other hand, when general relativity is used, there should be an independent method for estimating its accuracy. Without that, its results are uncertain.

The current book presents a more accurate description of physical reality. This description may or may not be as useful as general relativity. Time will tell.

ADDENDUM

BLACK HOLES?

In the earlier portion of this book it was noted that a probe can hit either another particle or another probe. The former hit causes the probe to end its cycle and return to its core. The latter hit generates a tiny bit of gravitational attraction between the two cores responsible for the two probes involved in the hit.

A black hole is presumably a star so massive that it exerts a gravitational force on the light leaving the star, a force strong enough to stop the light from leaving the star.

This latter idea is not reasonable. The force of gravity is only applied by probes, and it is only applied by one probe to one other probe at a time. The entire mass of the star is never exerted on a single probe at any time. Consequently, no matter how massive the star. it will always emit light.

But what about the possibility that the probe from the star will hit another particle, and thus be stopped? Now where did that other particle come from? It is presumed that the initial probe originated on the surface of the star. If the probe is flying off into the space surrounding the star, there should be very few such particles in the space near the star. It is hard to believe that there would be enough such particles to stop all of the vast number of probes emitted by the star.

Hence it seems reasonable to conclude that black holes do not exist.

Of course, supermassive stars do exist, and the implication of the above paragraphs is that they shine just as other stars do.

www.ingramcontent.com/pod-product-compliance
Lightning Source LLC
Chambersburg PA
CBHW081024170526
45158CB00010B/3145